Housing
119

换房子
House Sitting

Gunter Pauli

[比] 冈特·鲍利　著

[哥伦] 凯瑟琳娜·巴赫　绘

唐继荣　译

上海远东出版社

丛书编委会

主　任：田成川

副主任：闫世东　林　玉

委　员：李原原　祝真旭　曾红鹰　靳增江　史国鹏

　　　　梁雅丽　孟小红　郑循如　陈　卫　任泽林

　　　　薛　梅　朱智翔　柳志清　冯　缨　齐晓江

　　　　朱习文　毕春萍　彭　勇

特别感谢以下热心人士对童书工作的支持：

匡志强　宋小华　解　东　厉　云　李　婧　庞英元

李　阳　梁婧婧　刘　丹　冯家宝　熊彩虹　罗淑怡

旷　婉　王靖雯　廖清州　王怡然　王　征　邵　杰

陈强林　陈　果　罗　佳　闫　艳　谢　露　张修博

陈梦竹　刘　灿　李　丹　郭　雯　戴　虹

目录

Contents

一只寄居蟹正在海滩上找寻贝壳，用来做他的新家。

"这一定是你第五次想搬进一所新房子了吧。"一只海贝评说道。

"第五次？实际上，这将会是我第十个家。"螃蟹说。

"你为什么不能只找一处房子好好待着，何必老是四处寻找其他生物的房子？这可不是一种好的生活方式。"

A hermit crab is searching the beach for a shell that will become his new home.

"This must be the fifth time you want to move into a new house," comments the mollusc.

"Fifth? In fact, this will be my tenth home," Crab says.

"Why don't you just get one home and stay in it instead of always being on the lookout for other creatures' homes? This is no way to live your life."

一只寄居蟹正在海滩上找寻贝壳……

A hermit crab is searching the beach for a shell …

……长出一个壳，一层又一层。

... grow a shell, layer-by-layer.

"我没有脊骨，我需要一个外壳，你懂的。"

"那就造一座适合你的房子，也为将来长大留出地方。"海贝建议道。

"看呐，你说起来倒容易。"

"是你把生活搞艰难了。只需要将蛋白质与钙混合，就能长出一个壳，像我这样做——一层又一层。当你需要的时候，让它长得更大点。"

"I need a shell as I don't have a backbone, you see."

"So build something that fits you, and plan for growth," Mollusc suggests.

"Look, it is easy for you to say."

"You make life difficult. Just mix protein with calcium and grow a shell, like I do – layer-by-layer. Have it grow bigger when you need it to."

"你在表演魔法！多希望我也能做到，因为越来越多可能成为我的家的贝壳被人们拿走了。人们把这些贝壳做成工具、珠宝、武器、钱币、肥料，甚至用来制作乐器。"

　　"我喜欢长笛的声音，而用我们的壳发出的声音的节奏很有趣。我很高兴人们也喜欢它。"

　　"这些都是好的方面，但人们将所有贝壳都拿走了，我该住哪里呢？"寄居蟹想知道答案。

"You are performing magic! I wish I could, as more and more of my potential homes are taken away by people who use them as tools, jewellery, weapons, money, fertilizer and even to make music with."

"I love the sound of the flute, and the interesting rhythms of the sounds produced with our shells. I am pleased that people also like it."

"That is all good and well, but where am I going to live when people take all the shells away?" Crab wants to know.

……人们把这些贝壳做成工具、珠宝……

... people who use them as tools, jewellery ...

......贝壳同样是鱼类用来产卵的巢穴......

... shells are also nests for fish to lay their eggs ...

"我的朋友，这不只是关系到你和你的新家。人们从海底采集的贝壳同样是鱼类用来产卵的巢穴，也是海绵动物生长的基地。"

"那么，我们就不要再浪费时间讨论它了。我得去找一所房子。现在的这个家实在太小了，我住着不舒服。但是没有了它的保护，我可能随时都会被吃掉。"

"你需要一所房子，要足够大，有危险时能躲进去，又要足够轻，可以背负着到处走。"

"It is not just all about you and your new home, my friend. These shells that people collect from the ocean floor are also nests for fish to lay their eggs, and a rich base for sponges to grow on."

"So let's not waste time talking about it any longer. Let me go and find a home. The last one really is now far too small for comfort, and without it for protection I could be eaten at any time."

"You need a place that is large enough to hide in when there is danger, and light enough to carry on your back."

"对极了！你知道，我的朋友龙虾和螃蟹因为有坚硬的腹部所以不需要庇护所，只有像章鱼那样强健、巨大的嘴才能碾碎他们。但是我的腹部柔软，抵挡不了其他动物的攻击。"

"现在我懂了。随着你长大，你背着的房子就变得太小，住不下了，所以你需要找到新房子。"

"没错。你知道我们是群居动物吗？"

"群居？少骗我了！一个贝壳里能住多少只寄居蟹？"

"Exactly! You know, my lobster and crab friends do not need a shelter as they have a hard belly. Only tough, big mouths like those of the octopus can crush them. But I have a soft belly."

"Now I understand. When you grow bigger and the house you carry becomes too small, you need to find a new house."

"That is right. You know that we are social animals?"

"Social? You could have fooled me! How many crabs live in one shell?"

……章鱼那样巨大的嘴……

... big mouths like those of the octopus ...

······我们会排队。

... we stand in line.

"只有我一个！但是，当我们需要搬进更大的贝壳时，我们会排队。"

"是为了第一个得到新的可供利用的贝壳吗？"

"不是的！虽然我们的确会争斗，但这类事情很少发生。我们排队的目的是：当我得到更大的贝壳时，排在我后面那个家伙将拥有我留下的贝壳，而排在更后面的那个将得到更小的贝壳。因此，哪怕只有一个新的贝壳可供利用，我们这样做能让数十只寄居蟹立即根据体型更换新家。"

"Just me! But when it is time for us to move into bigger shells, we stand in line."

"To be the first to get a newly available shell?"

"No, we seldom fight, although it does happen. We stand in line so that when I get my larger shell, the next one in line gets my shell, and the next in line gets that smaller shell, and so dozens of us can change homes at once – according to our size – when only one new shell becomes available."

"所以，你们是回收利用的行家！"

"是的！我们回收那些因为原有住户长大而住不下的贝壳。当我们长大，无法在原有贝壳里居住时，就把它们留给体型更小的寄居蟹。"

"你们住进去的最大的贝壳有多大？"

"So you are master recyclers!"
"Yes, we recycle the shells of those who have outgrown them. When we outgrow them, we pass them on to smaller crabs."
"And what is the biggest shell you can get into?"

......留给体型更小的寄居蟹。

... pass them on to smaller crabs.

······像椰子那样大的贝壳!

... a shell the size of a coconut!

"噢！我们不断生长，直到搬进像椰子那样大的贝壳！"

"那么大？哇！"

"不仅如此——我们最长能活过70年。"

"太神奇了！请告诉我，谁是你们最好的朋友？"

"Ah, we can grow until we fit into a shell the size of a coconut!"

"That big? Wow!"

"And that is not all – we can live for up to seventy years."

"Amazing. And tell me, who are your best friends?"

"当然是你们海贝啦，因为你们给我们提供安家的居所。海葵排第二，因为他们把捕食者吓跑，然后捡拾我们剩下的食物作为回报。"

"你们不仅能长大、长寿，还是快乐的隐士——有很多朋友围绕的隐士！"

……这仅仅是开始！……

"You molluscs, of course, as you provide us with homes. After that, it is the sea anemones. They scare away predators and in return pick up our left-over food."

"So you not only get to grow big and old, you are happy hermits – with many friends around!"

... AND IT HAS ONLY JUST BEGUN! ...

……这仅仅是开始！……

... AND IT HAS ONLY JUST BEGUN! ...

Did You Know?

你知道吗？

The land-based hermit crab known as the coconut crab, is one of the world's largest invertebrates.

一种被称为椰子蟹的陆寄居蟹，是世界上最大的无脊椎动物之一。

A very large percentage of all animals (about 95%) have no backbone or spine. This includes corals, spiders, crabs, jellyfish, sponges and worms.

绝大部分动物种类（约95%）没有脊椎或脊柱，包括珊瑚虫、蜘蛛、螃蟹、海蜇、海绵动物和蠕虫。

Hermit crabs queue up from the biggest to the smallest, waiting for a bigger one to shed its shell. This causes a chain reaction of shell changes, each moving in to a shell the next size.

寄居蟹从最大到最小依次排队等候大的个体蜕壳。这就产生了一种换壳链式反应，即每个个体移居到相邻尺寸的贝壳中。

90%

10%

Mollusc shells have a spiral. More than 90% of mollusc species have right-handed spirals, and a minority has left-handed spirals. A few mollusc species have both left and right spiralling shells.

软体动物的外壳有螺纹。超过 90% 的软体动物物种螺纹为右旋，少量软体动物为左旋。有一些软体动物的壳既有左旋螺纹也有右旋螺纹。

拥有约 8.5 万种成员的软
体动物大家族是最大的无脊椎
动物家族，它代表 23% 的已知
并被命名的海洋生物。科学家
预期软体动物种数可增加到近
12 万种。

The mollusc family, with about 85,000 known members, is the largest family of invertebrate animals and represent 23% of all known and named marine organisms. Scientists expect that this will increase to around 120,000 species.

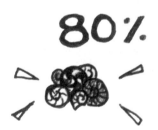

中国的软体动物捕捞量占
全球的 80%，每年网捞 1100
万吨。法国的软体动物捕捞量
位居全球第二。

China accounts for 80% of the global mollusc catch, netting 11 million tons per year. France has the second biggest catch.

过去，贝壳主要在环印度洋地区被用作钱币，使用区域从马尔代夫群岛、斯里兰卡、加里曼丹岛、东印度群岛，到沿非洲海岸向南至莫桑比克一带。

Shells were used as money mainly around the Indian Ocean, from the Maldives Islands, Sri Lanka, Borneo, and the East Indian Islands, and along the African coast as far south as Mozambique.

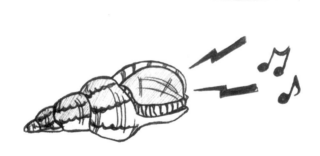

大法螺的贝壳一直被用作小号。大法螺在日本被称为 horagai，在马尔代夫被称为 sangu，而新西兰的毛利人称之为 pūtātara。

The giant triton shell has been used as a trumpet. In Japan it is known as a horagai, in the Maldives as a sangu and in New Zealand the Maoris call it a pūtātara.

Think About It

想一想

Would you like to change your home as you grow, taking over someone else's?

你愿意随着长大而改变住处，去接手别人的住处吗？

Can you imagine life without a spine? What would you still be able to do?

你能想象没有脊柱的生活吗？在这种情况下，你还能做什么事情？

Now, what about a house that grows as you grow, without ever having to change the basic structure, just by you continuing to add on to it?

如果有一套随着你长大而长大，只需不断增补材料，甚至不要改变基本结构的房子，你觉得怎么样？

Would you pick up a shell on the beach that could be the home of a hermit crab? Would you feel bad about depriving the crab of a new home?

你会在海滩捡拾那些可能被寄居蟹当作家的贝壳吗？你会为夺走蟹的新家而感到难过吗？

How many shell shapes can you recognise? Shells have the most diverse shapes and yet there are some basic characteristics that apply to all. Study the shapes and then describe at least six different ones. Now get acquainted with the geometry of their shapes. Discuss the advantages and the benefits of each of these shapes and correlate this with the area where these shells are formed: calm or stormy water, abundance of sun and oxygen or not. Think logically about the environment in which each shell is shaped and what advantages it holds for the molluscs inside.

你能识别多少种贝壳形状？贝壳的形状极其多样化，但所有的贝壳也具备一些共同的基本特性。研究一下贝壳的形状，并描述至少6种不同的形状。熟悉了贝壳形状的几何学之后，讨论一下每一种形状的优势和好处，并与这些贝壳形成的地区关联起来：静止或波涛汹涌的水域，阳光、氧气充足或与此相反。从逻辑上思考每个贝壳形成的环境以及它对里面的软体动物有什么好处。

学科知识
Academic Knowledge

生物学	共生关系；无脊椎动物是没有脊椎或脊柱的动物；腹足类是能为寄居蟹提供外骨骼的软体动物；软体动物是取食藻类的草食性动物，或为滤食性动物；海藻学是研究海藻的科学。
化 学	贝壳的钙质中心层由沉积在一种被称为贝壳素中的有机基质的碳酸钙形成；贝壳的外层耐摩擦；贝壳的内层由珍珠母组成；把贝壳融入水中会升高水体的pH值；从尿素中分解出的铵离子会提高pH值，促进碳酸钙的沉积。
物 理	一些蜗牛（或螺类）具有能关闭贝壳的密封口，为其提供全面保护；贝壳的结构是外部为一层棱镜层，内部为珍珠状片层。
工程学	软体动物的外壳在其整个生命周期稳定生长，通过向边缘增加碳酸钙，使贝壳以一种螺旋形增长的方式变长和变宽。
经济学	软体动物可作为生物指示器来监测水体环境在恢复力和生产力上的健康程度；海洋酸化影响超过十亿以海产品为主要蛋白质来源的人的生计。
伦理学	为什么当现存的结构还能使用时，还要消耗很多原始材料来构建新的结构？
历 史	软体动物最早出现于寒武纪；从公元前2000多年的商朝直到周朝，中国把子安贝用作钱币；玛雅人喜爱太平洋海菊蛤（又称刺牡蛎），它具有隐含的红色，象征日出和重生；从工业革命开始，海洋表水的pH值已降低了0.1，这一数值变化表明酸度有高达30%的改变。
地 理	贝壳过去主要在环印度洋地区被用作钱币。
数 学	软体动物外壳围绕主轴的每轮生长率、曲线的形状、生长曲线的平移率和生长随动物达到成熟期的变化都可以建立数学模型；鹦鹉螺的外壳与黄金比例ϕ（0.618）相关。
生活方式	软体动物已成为人类重要的健康食物来源；通常可食的软体动物包括章鱼、鱿鱼、海螺、牡蛎和扇贝。
社会学	软体动物为人类社会提供奢侈品，尤其是珍珠、珍珠母、提尔紫染料和海藻纤维。
心理学	中国人认为家里放置贝壳会增添平安富贵的感觉，并如同盾牌一样为家庭提供保护，寓意一帆风顺和永不分离。
系统论	软体动物是碳酸钙的主要的沉积者，它们的溶解增加了水体的pH值，为生命在海洋中的存在创造条件；软体动物被分为10个纲，其中2个纲现已完全灭绝；将各种螺类引入新的区域已经对一些自然生态系统造成严重的损害。

情感智慧
Emotional Intelligence

海 贝

海贝善于观察, 他注意到寄居蟹在搜寻房子, 但好奇寄居蟹为什么不一劳永逸地建造一座房子, 并认为他没必要把日子过得这样艰难。海贝欣赏用贝壳产生的音乐, 但并不清楚贝壳采集对寄居蟹的日常生活所带来的冲击。然而, 他的确意识到贝壳对鱼类和海绵动物的重要性。他变得更设身处地, 对寄居蟹需要的房子尺寸和重量进行反思。海贝一开始并不理解寄居蟹不断找寻新房子的行为, 一旦明白寄居蟹发现一处新家的真正过程和这将如何影响其社会生活, 他的质疑转变为赞赏, 他渴望去理解和知晓更多。他得出结论: 寄居蟹过着一种幸福的生活。

寄居蟹

寄居蟹感到有压力, 因为他的体型超过了他的房子。面对批评和意见, 他努力辩解为何他需要一座新房子。当海贝提出不可能实现的建议时, 寄居蟹并不认同。他通过向海贝表达钦佩来重回原来的交谈, 同时抱怨采集贝壳的人类偷走了他潜在的家。寄居蟹是急性子, 不打算浪费更多时间争辩。他急需一个家并找到解决方案。当海贝看上去准备好帮他找到解决方案时, 寄居蟹进一步解释他的困境, 并把自己描述为具有社会结构的动物。面对海贝的质疑, 寄居蟹耐心地娓娓道来, 最终获得海贝的钦佩。

艺术
The Arts

为你收集到的海贝涂色, 给源于自然的精致而多样的产品增添一份个人色彩。水彩颜料能让贝壳表面的某些自然特性透出来, 而用丙烯酸颜料则会丢失这些自然特性。从贝壳生存的背景开始画, 然后绘制一些简单的形状, 例如有眼睛和鼻子的脸部。将画作密封好, 确保染料不脱落。在你创造美的过程中, 一定要检查染料和密封器的成分, 避免使用含有重金属的产品。

思维拓展
Systems: Making the Connections

自然界提供了如此多的利用材料效率的例子。首先是寄居蟹、软体动物、海葵和苔藓动物的共生关系，共生为不同的物种提供了食物、居所和保护，而这是任何单个物种无法完全做到的。营养物和物质的传递导致海洋中碳酸钙的堆积。这种明显的材料效率得到另一种形式的效率的补充，在经济学上可解释为：通过延长现有结构（贝壳）的生命周期更好地利用生产资料（或资本货物）。这使得研究海洋生物（寄居蟹为海陆两栖）变得如此有趣。纵观历史，贝壳早已为人类所用，如用于房屋装饰，但也被赋予海洋酸度调节器的特殊价值。生命从海洋开始，这里的pH值已经保持长达2 000万年的稳定。由于现代生活排放过量的二氧化碳，海洋和海贝的作用正从适度向动态不平衡转变。二氧化碳溶解于海水后转变为碳酸，导致海洋酸化。另一方面，二氧化碳也提供碳，它与已经溶解在海洋中的钙结合形成碳酸钙。这在丰富的海洋生物多样性中引发一个自然选择过程：一些物种难以应对升高的二氧化碳浓度，生存环境恶化，导致支撑生命的结构溶解，将碳释放到水体中。然而，当海水中含有更多的二氧化碳时，甲壳类动物看上去长得更重，并储存更多的碳。藻类和海草作为植物也可能受益，因为它们需要二氧化碳才能生存。海洋中酸度升高对海珊瑚、牡蛎和蛤类有影响，有可能把养活千百万人的整个食物链置于险境。只有人类显著降低他们的温室气体排放，这个可能有不确定结果的复杂系统才能恢复其动态平衡。

动手能力
Capacity to Implement

邀请朋友来帮你精选不同形状和尺寸的贝壳。研究利用这些贝壳创作音乐的可能性。某些贝壳可能发出哨音，而其他一些则可能成为打击乐器的理想选择。对所有可能由贝壳产生的声音进行试验。感受各种各样的声音，随着贝壳变大，声音也从尖锐的哨声转为更深沉的基调，甚至可能产生回声。你能把不同的贝壳排列组合，并创造一些和声吗？

故事灵感来自
This Fable Is Inspired by

西尔维娅·厄尔
Sylvia Earle

生于 1935 年的西尔维娅·厄尔是一位海洋生物学家和探险家。西尔维娅在年轻时随家庭移居佛罗里达州，在那里她发展出对户外生活的巨大热情。她先后于 1955 年和 1966 年从杜克大学获得理学硕士和藻类学博士学位。她是"可持续海洋探索"机构的领导人和"谷歌地球"海洋咨询委员会主席，而且用一套单人常压潜水服在开放大洋创造了 381 米的下潜深度纪录。为了探索和保护海洋，西尔维娅还创立了名为"蓝色使命"的非营利组织，1998 年，美国《时代》杂志称她为第一位"地球英雄"（Hero for the Planet）。从那时起，她成为美国国家地理学会的驻会探险家。

图书在版编目(CIP)数据

冈特生态童书.第四辑:修订版:全36册:汉英对照 /
(比)冈特·鲍利著;(哥伦)凯瑟琳娜·巴赫绘;
何家振等译.—上海:上海远东出版社,2023
书名原文:Gunter's Fables
ISBN 978-7-5476-1931-5

Ⅰ.①冈… Ⅱ.①冈… ②凯… ③何… Ⅲ.①生态环
境–环境保护–儿童读物—汉、英 Ⅳ.①X171.1-49

中国国家版本馆CIP数据核字(2023)第120983号
著作权合同登记号图字09-2023-0612号

策　　划　张　蓉
责任编辑　张君钦
封面设计　魏　来李　廉

冈特生态童书
换房子
[比]冈特·鲍利　著
[哥伦]凯瑟琳娜·巴赫　绘
唐继荣　译

记得要和身边的小朋友分享环保知识哦!
八喜冰淇淋祝你成为环保小使者!